The Adventures of Sojourner

The mission to Mars that thrilled the world

M I K A Y A   P R E S S

NEW YORK

For Eric and Noah

Acknowledgements

The author gratefully acknowledges the assistance of:
Donna Shirley
Manager of the Mars Exploration Program, Jet Propulsion Laboratory

Guy Beutelschies, Brian Cooper, Matthew Golombek, Jennifer Harris,
Bridget Landry, Rob Manning, Jacob Matijevic and Brian Muirhead
Members of the Mars Pathfinder Mission Team

Diane Ainsworth
Public Information Office, Jet Propulsion Laboratory

Editors: *Stuart Waldman and Elizabeth Mann*

Image Editing and Research: *Stuart Waldman*

Design: *Lesley Ehlers Design*

Published by Mikaya Press Inc.
No part of this publication may be reproduced in whole or in part or stored in a retrieval system,
or transmitted in any form or by any means, electronic, mechanical, photocopying,recording or otherwise,
without written permission of the publisher. All rights reserved. For information regarding permission, write to:
Mikaya Press Inc.,12 Bedford Street, New York, N.Y.10014.
Distributed in North America by: Firefly Books Ltd., 3680 Victoria Park Ave., Willowdale, Ontario, M2H3KI

Library of Congress Cataloging -in-publication Data

Wunsch, Susi Trautmann.
The adventures of Sojourner : the mission to Mars that thrilled
the world / by Susi Trautmann Wunsch.
p.   cm.
Summary : Tells the story of the mission that placed the Sojourner
remote-control rover on Mars on July 4, 1997.
ISBN 0-9650493-5-3. — ISBN 0-9650493-6-1 (pbk .)
1. Mars (Planet)—Exploration—Juvenile literature.
2. Space flight to Mars—Juvenile literature. 3. Sojourner (Spacecraft)—
Juvenile literature. [1. Mars (Planet)—Exploration. 2. Space
flight to Mars. 3. Sojourner (Spacecraft)] 1. Title.
QB641. W86   1998
919. 9' 2304—dc21   98-7660
                                            CIP
                                            AC

Printed in Singapore

of
Sojourner

The mission to Mars that thrilled the world

By Susi Trautmann Wunsch

MIKAYA PRESS

NEW YORK

On September 1, 1993, Michael Dean frantically tapped out signals on his computer keyboard at the Jet Propulsion Laboratory in Pasadena, California. "Speak to me," thought the young flight controller. "Give me something to lock onto."

Day and night for almost a week, Michael had struggled to restore communications with NASA's Mars Observer spacecraft. (NASA is the United States government agency in charge of space exploration.) The $1 billion Observer carried cameras and scientific equipment to make detailed maps of the planet. The spacecraft had taken 8 years to build and had traveled through space for more than 11 months. Then, only 3 days before it was to begin orbiting Mars, Observer had suddenly quit talking to Earth.

Michael commanded Observer to turn on its backup radio. He conferred for hours with engineers who took apart a replica of the spacecraft, piece by piece, to try to determine what had gone wrong. They found nothing.

The flight team huddled anxiously around Michael's screen waiting for an electronic blip from the spacecraft.

Still there was nothing. Mars Observer was lost in space.

Observer was the 19th mission to Mars since the former Soviet Union launched the first one in 1960. But this mysterious planet fascinated people long before anyone ever dreamed of traveling into outer space. The Egyptians called it the Red One, the Babylonians called it the Star of Death. The Romans named it Mars after their god of war because its reddish color reminded them of blood.

With the invention of the telescope, astronomers began to learn about this cold, dry planet and its two small, potato-shaped moons. Some thought they saw canals on the surface of Mars, indicating the presence of intelligent life.

By the early 1900s, more powerful telescopes revealed no evidence of canals, but the idea that life might exist on Mars had taken hold. It inspired tales of Martian invaders and interplanetary heroes.

There were good scientific reasons to look for life on Mars. Mars is a lot like Earth. It has an atmosphere, polar caps, and there may even be water beneath its surface. It's the only other planet in this solar system with the right stuff to support life. It's also the most likely place to build colonies where humans could live in the future.

Mars today.

Mars as it might
have looked billions
of years ago, with
water and a thicker
atmosphere.

The earliest missions to fly past Mars in the 1960s — Mariners 4, 6 and 7 — sent back photos of a disappointingly dead, moon-like surface.

Between 1964 and 1976, the United States launched 6 successful missions to Mars. Some flew by the planet and took photographs, some orbited and some even landed and tested the soil. The evidence they gathered led most scientists to conclude that, right now, life probably does not exist on the surface of Mars. But that doesn't mean it never did.

Photographs suggest that water might have flowed across Mars at one time. Scientists reason that where there had been liquid water, there must have been a thicker atmosphere and a warmer climate — just the right conditions for primitive life to evolve. Could it have happened long ago on Mars?

To find out, scientists had to study the rocks. On Earth, analysis of rocks has revealed much about the age of our planet and how it evolved. The chemical elements inside tell geologists what forces formed them: water, volcanoes or the slamming together of floating land masses. Sometimes the rocks even preserve signs of past life in fossils. As on Earth, the story of Mars would unfold through its geology.

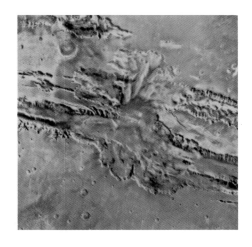

In the 1970s, photos from Mariner 9 and the Viking missions revealed a vast canyon called Valles Marineris along the Martian equator. Channels at its eastern end were among the first clues that water might have once flowed on Mars.

Robby was an early design for a Mars rover that could travel hundreds of miles collecting rocks. It had to be large just to hold the bulky computers that would help navigate it.

But there was a problem. Although spaceships could land on Mars, how could they reach the rocks once they arrived? A mobile robot, or rover, was the answer. Controlled long-distance from Earth, a rover could be steered across the Martian terrain to the rocks the geologists wanted to study.

In the late 1980s, Donna Shirley, an engineer at the Jet Propulsion Laboratory who had dreamed since childhood of journeying to Mars, had the job of overseeing a variety of plans for rovers. One called for landing a rover the size of a pickup truck on Mars. It would have been

able to heave rock samples into a spacecraft that would shuttle them back to Earth. The mission's $10 billion price tag proved too high. The proposal was scrapped in 1991.

Then, in 1993, came the loss of the Mars Observer. A tiny fuel leak triggered an explosion that hurled the costly spacecraft far beyond the reach of flight controllers before it even entered Mars orbit.

With the loss of the $1 billion mission, NASA realized that interplanetary space exploration had become too complicated and expensive. The space agency had to try a different approach. "Faster, better, cheaper" became NASA's new motto.

Donna Shirley and her team were in the right place at the right time. The same advances in computer technology and miniaturization that brought laptops and cell phones into people's homes now made simple, inexpensive interplanetary rovers possible.

Geologists might have liked to examine the large rock that dominates this Martian landscape, but the Viking 1 lander wasn't designed to move. It could only scoop up soil samples that were right next to it.

Apollo astronauts drove this jeep-like rover on the moon. The little Sojourner rover could have occupied its passenger seat.

By 1994, Donna's team was using the new technology to develop a microrover that would become the first mobile, remote control explorer of another planet. It would be called Sojourner (after Sojourner Truth, a crusader for the emancipation of slaves and for women's rights during the Civil War era). Sojourner would make her historic journey as part of the Mars Pathfinder mission.

The Pathfinder mission was scheduled for launch in December of 1996. The space program had never attempted anything quite like it. There would be just 3 years between development and liftoff. Only 20 to 30 people, compared with hundreds who worked on previous Mars missions, would operate it. Among them would be a new generation of young, enthusiastic scientists and engineers who were unafraid of taking risks. Many had been kids when astronauts first walked on the moon in 1969. Now it was their turn to stake out a new frontier.

Consistent with the "faster, better, cheaper" way of thinking, Pathfinder set out to demonstrate an efficient, inexpensive way to land on Mars. The intricate retrorocket landing system and all the testing that had been necessary to assure the gentle touchdown of earlier Mars landers would be too costly for this mission. Besides,

This painting shows how the Pathfinder lander would look with its air bags deflated, its solar panels fully opened and the Sojourner rover beginning its exploration of Mars.

a conventional lander that sat on legs could tip over easily on a rock-strewn landing site.

Mars Pathfinder flight system manager Brian Muirhead and the lander team took a daring approach. They designed a legless, 3-foot tall, pyramid-shaped lander that would plunge into the thin Martian atmosphere at tremendous speed. A parachute would slow its descent. Then a giant cluster of air bags would inflate around the lander just before it slammed onto the surface, cushioning its impact.

Once on Mars, the air bags would deflate, exposing the lander. Its solar panel sides would open like flower petals to capture energy from the sun and to reveal the lander's cargo: a camera, antennas, weather apparatus and the Sojourner rover. After the rover team studied photos of the landing site to be sure it was safe, Sojourner would be commanded to drive off the lander and begin to study the Martian rocks and soil with her scientific instruments.

In the test area called the Mars Yard, engineers tried out the air bags that would shield Pathfinder during its risky bounce landing. To stand up to jagged rocks, the air bags were made of 5 layers of Vectran, a material used for bulletproof vests.

Rover manager Dr. Jacob Matijevic and his team set to work to build Sojourner. To keep down the costs of flying her to Mars, Sojourner had to be bantam-weight and -size. The rover measured only 2 feet long, 1 1/2 feet wide and 1 foot tall when extended to her full height. She was constructed of lightweight, yet strong, aluminum. Fully loaded, Sojourner weighed just over 23 pounds.

Sojourner's main scientific instrument was called an APXS, which stands for alpha proton X-ray spectrometer. The APXS would bombard a rock with radioactive particles. By "reading" what bounced back, the APXS could identify the chemical composition of the rock. Each reading would take about 10 hours.

Sojourner's exploration of Mars would require less electricity than it takes to light a refrigerator bulb. A solar panel would power Sojourner's daytime operations. She would use ordinary flashlight batteries for night duty.

The rover team adapted store-bought radio modems for use on Mars. They designed and built the antennas themselves.

The severe cold—with temperatures falling as low as -127 degrees Fahrenheit at the landing site— could damage mechanical and electronic parts. With no money to develop special cold-resistant motors, ordinary motors were placed in a box that was lined with an almost weightless insulation. Three tiny radioactive heaters would keep the box warm.

Spikes on Sojourner's 6 wheels would help her to plow through soft soil and to gain a foothold on rocks. The front and rear wheels could be steered separately so that Sojourner could turn in place. Tilt sensors, like those on a pinball machine, would prevent her from flipping over and becoming stranded on Mars like a turtle on its back.

Brian Cooper at the rover control workstation. The round object in his right hand is a joystick to guide the image of the rover on the screen.

Sojourner driver Brian Cooper, a video game master since childhood, developed the software for the highly sophisticated driving system. Although Sojourner looked like a remote control toy car, steering her across Mars would be considerably more complicated. It would take at least 10 minutes for signals to travel across the 119 million miles that separated Mars from Earth. Sojourner could roll over a cliff before an order to halt reached her.

Brian had to plan the best and safest route in advance. He would sit at a computer console with a set of battery operated goggles over his eyes. Through the goggles, he would see 3-D images of the landing site taken by the camera on the Pathfinder lander. Using a joystick, Brian would "steer" a rover-shaped cursor to a particular rock or other feature. This destination, and the instructions for getting there, would be radioed to Sojourner. To assure her safety, the rover would move very slowly. Her top cruising speed would be only 2 feet per minute.

A view of Brian's computer screen. A model of the rover rests atop a small hill in a test area. The green dart icons mark the path that the rover is to follow around a large rock.

Laser beams would help Sojourner pick out the shapes of her intended targets and avoid dangerous obstacles.

Sojourner wouldn't be completely reliant on Brian. She would also be able to navigate independently. The rover would use her hazard avoidance system to shut down automatically in case of danger. Two forward-mounted camera "eyes" and 5 laser beams projecting from her forehead would enable Sojourner to sense obstacles and how far away they were from her. Depending upon the sizes and shapes of the hazards, Sojourner's computer "brain" would help her to decide whether to climb over them or steer clear.

Sojourner would pause frequently to evaluate the landscape. Her computer would update the distance that she had traveled by counting the number of turns of her wheels. A gyroscope, spinning like a top inside her, would keep the rover's course steady. Like a watchful parent, the Pathfinder lander would listen for Sojourner's heartbeat, a radio signal that the rover sent out continuously during her surface operations.

When Jake and his team had finished building Sojourner, they proudly autographed a sheet of silver Mylar and affixed it to the little rover's belly. After a few more months of testing, Pathfinder and Sojourner left the Jet Propulsion Laboratory in Pasadena to travel to the launch site: the Kennedy Space Center at Cape Canaveral, Florida.

During the week of August 12, 1996, a truck caravan drove the Pathfinder lander from California to Florida. The little rover followed later, packed into the cargo hold of a regular commercial jet.

After 3 years of planning and development, all was ready for the Pathfinder lander and the Sojourner rover to begin their 7-month journey to Mars.

At Cape Canaveral, Sojourner was collapsed to a height of only 7 inches and lashed with cables to the inside of one of the Pathfinder petals. Pathfinder, in turn, was strapped inside a hard, capsule-shaped "aeroshell" which was covered by a heat shield. The heat shield would protect the capsule from the scorching friction of high-speed entry into the Martian atmosphere.

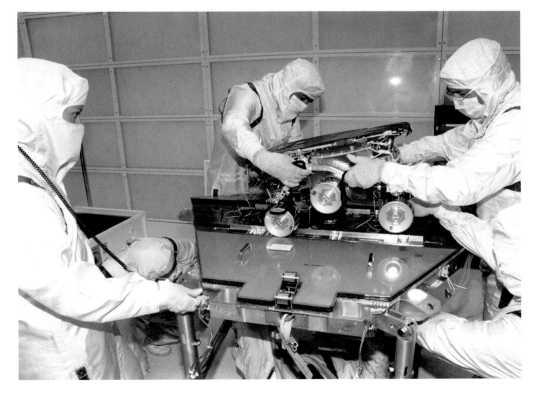

Technicians gingerly lift the Sojourner rover onto the lander's petal for its long journey to Mars. They are in the "clean room." Just as people on Earth worry about microbes from outer space infecting us, NASA follows strict rules to protect other planets from Earth's microorganisms.

Heat Shield ⎤
               ⎦ Capsule
Aeroshell ⎦

Cruise Stage

Third Stage

With the Pathfinder lander inside, the capsule is attached to the third stage of the Delta II launch vehicle. (The aluminum collar going around the base is for protection during the trip to the launch pad. It will be removed before launch.)

Then the capsule was plugged into the cruise stage, which would take Pathfinder to Mars. It was fitted with all the necessary gear for interplanetary flight: small thruster rockets, navigation equipment, solar panels for power and an antenna to keep in touch with Earth.

A Delta II 7925 would be used to launch the spacecraft on its flight to Mars. The Delta II is a small and inexpensive launch vehicle. Each of its three sections, or stages, propels the spacecraft until that stage's fuel is gone. Then it falls away. Every time one stage is shed, the stages that remain get lighter and less fuel is needed to propel the spacecraft higher and faster.

The early December launch date was picked for a special reason. Earth and Mars both orbit the sun, but the orbit of Mars is much wider than that of Earth. Only about every 26 months are the 2 planets lined up so that the minimum amount of fuel would be needed during launch. The Pathfinder team also timed the launch so that the spacecraft would reach Mars on America's birthday, the 4th of July, 1997.

Nose Cone

Heat Shield

Aeroshell

Cruise Stage

Third Stage

On top of the launch tower, the third stage is mated with the other two stages of the Delta II launch vehicle. The technician at the lower right is moving part of the nose cone into place. The nose cone shields the spacecraft as the rocket speeds through Earth's lower atmosphere and then it falls off.

On December 4, Donna Shirley impatiently awaited Pathfinder's launch. The blue and white rocket stood out like a monument in the glow of spotlights. The nose cone at its tip covered Pathfinder like a protective shield.

The launch had already been delayed twice, once because of bad weather and again, with only 30 seconds to go in the countdown, because of a computer malfunction. "Let's go!" thought Donna.

In the Pathfinder control center, launch conductor Guy Beutelschies eyed an electronic countdown clock and a TV monitor that displayed different views of launch pad 17B.

"Communications. Ready for launch?"

"Communications is go for launch," came the reply.

"Systems analysis. Ready for launch?"

"Systems analysis is go for launch."

"Spacecraft is go for launch," Guy reported to the NASA launch coordinator.

As Pathfinder moved closer to launch, the Deep Space Network prepared for its part of the mission. The giant antennas of the Deep Space Network, including this one in Canberra, Australia, are NASA's interplanetary listening posts. Three stations are located at key points around the globe to provide two-way communications with spacecraft such as Pathfinder. The Deep Space Network would locate Pathfinder's position during flight, transmit commands from Earth and receive images and data from Mars.

Pathfinder would land on an ancient Martian flood plain known as Ares Vallis. Scientists think that waters equal to more than 1,000 times the flow of the Amazon River might have whooshed through the site, depositing a wide variety of rocks. At least that's what the geologists hoped to find.

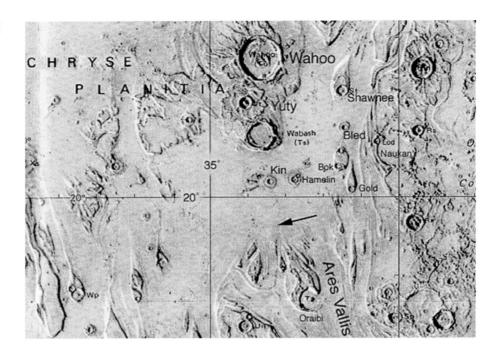

The flight team monitoring the launch back at the Jet Propulsion Laboratory in Pasadena also reported that it was ready to go.

The seconds on the clock ticked by. Over his headset, Guy could hear: "...three, two, one. Ignition. We have ignition. Liftoff." With a mouse click to the launch button, the engines ignited and the rocket lifted off at 1:58 a.m. The blast momentarily turned night into day. Donna Shirley could feel the earth shake beneath her as the launch vehicle rose vertically for the first 2 minutes, its progress slow because of the weight of the first-stage fuel. Then the rocket arched gracefully into the clear night sky. To Donna, Mars appeared as a red speck to the left of a crescent moon that seemed to smile on the mission.

For 4 minutes, the 9 strap-on rockets around the first stage helped to boost the spacecraft up through the densest part of Earth's atmosphere. Then they fell away into the Atlantic Ocean.

The launch vehicle follows the curvature of the Earth. A few minutes after lift-off, the bright fireball faded to a pinprick, and then the spacecraft was beyond sight of observers on the ground.

The smaller second stage lifted the spacecraft higher and faster into low Earth orbit, before it fell away and burned up on reentry into Earth's atmosphere.

Like a pitcher letting go of a fastball, the third-stage booster hurled the spacecraft beyond the reach of Earth's gravity and toward Mars. Then the third stage separated. All that remained of the 123-foot prelaunch stack was the Pathfinder capsule attached to its cruise stage.

The spacecraft would "coast" toward the landing site at 12,000 miles per hour, unaided by further rocket power because of a certain law of physics that was discovered by Sir Isaac Newton more than 300 years ago. The law says that an object placed in motion stays in motion in an airless place like outer space.

Pathfinder was aimed at a target approximately 125 miles long and 62 miles wide that was hundreds of millions of miles away.

To understand how exact the course calculations had to be, think of tossing a basketball in Houston, Texas into a hoop in Los Angeles, California. Now, imagine that the basket is moving. For, as the spacecraft followed an arcing path defined by the sun's gravity, Mars orbited the sun too, but in a wider loop. On landing day, the spacecraft and planet would have to meet at a certain point in space.

From a computer console presided over by a smiling toy spaceman, chief navigator Pieter Kallemeyn kept track of Pathfinder and recommended adjustments to its course. During Pathfinder's long journey to Mars, the flight team would have 5 opportunities to adjust the spacecraft's course by firing the small thruster rockets on the cruise stage for a few seconds at a time.

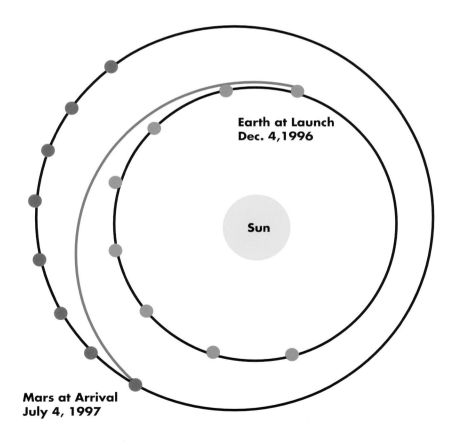

Earth at Launch
Dec. 4, 1996

Sun

Mars at Arrival
July 4, 1997

The Pathfinder spacecraft was aimed at the point in space where Mars would be on July 4, 1997.

Gremlin Gruel used an identical twin of Sojourner called Marie Curie while rehearsing in the test bed.

While Pathfinder flew, the people on the ground rehearsed what would happen after the landing. An engineer named David Gruel tried to predict what obstacles Mars might throw into the path of the Sojourner rover. David's friends had nicknamed him Gremlin Gruel. (Gremlins are mischievous beings that are thought to cause mechanical failures in ships and airplanes.)

Every Sunday night, Gremlin would enter a test bed, a large room with a sand-covered floor. He would close the blinds behind him so that the other team members could not peek inside. Then he would take out his shovel and a bag of rocks and build obstacles around full-scale models of the Pathfinder lander and the Sojourner rover. He would place the lander at a bizarre tilt on a pile of rocks, or run the rover into a hillside.

On Monday morning, Gremlin would challenge mission engineers to solve the problem as if it had happened on Mars. The blinds remained shut. The engineers could only use the images and data that the lander and rover sent to Brian Cooper's computer outside the test bed. The solutions they came up with were fed into the computer program that would help Sojourner when she ran into similar problems on Mars.

The rover practices climbing over rocks up to 8 inches tall.

Brian and the engineers practiced many times in the test bed. Like musicians preparing for a performance, they gained confidence with each rehearsal.

In the Mission Operations room, glowing red numbers on an electronic scoreboard ticked off the days, hours, minutes and seconds until the landing.

The team grew anxious as landing day approached. They thought: Are we ready? Have we anticipated every problem? Have we forgotten anything? Chief flight engineer Rob Manning dreamed that he launched his dog Scooter to Mars, then realized too late that he had neglected to figure out how to bring him back home.

On the day before touchdown, Pathfinder entered the embrace of Mars gravity, picking up speed as it fell toward the landing site. A navigator checked to see whether a 5th and final course correction would be required. It was not. Pathfinder was right on target. It was July 4, 1997.

Pathfinder separates from the cruise stage
Time until landing: 35 minutes

Pathfinder Enters Mars Atmosphere
Altitude: 80.8 miles
Time until landing: 5 minutes

Parachute Opens
Altitude: 5.8 miles
Time until landing: 134 seconds

Heat Shield Separates
Time until Landing: 114 seconds

Pathfinder detached from its cruise stage and dove through the dark Martian sky at 16,700 miles per hour. The friction of entry into the atmosphere slowed the spacecraft to 828 miles per hour and caused the heat shield to glow red hot. Tiny glass balls embedded in the heat shield melted and emitted gases that acted like fire extinguishers.

If you had been standing on Mars that day, here is what you would have seen: Glowing like a meteor, Pathfinder would have blazed across the predawn Martian sky. The Earth would have been visible as a blue morning star in the background.

In the Mission Operations room, a tense crowd clustered around Rob Manning, who was monitoring Pathfinder's entry, descent and landing. They waited.

A signal alerted Rob that Pathfinder's parachute had opened, slowing the descent to 150 miles per hour. What Rob could not hear was the fireworks on Mars that Independence Day. The explosion that opened Pathfinder's parachute was among 42 that were needed to trigger each step of the most complex landing system that the Jet Propulsion Laboratory had ever built.

Pathfinder's heat shield blew off. The lander detached from its aeroshell and whizzed down a 65-foot long tether like a fireman descending a firehouse pole.

Ten seconds before impact, the air bags inflated. Three small braking rockets fired, their thrust yanking the parachute and aeroshell up and away from the lander. Pathfinder hung motionless in midair for an instant before plunging the final 70 feet to the ground.

**Entry, Descent and Landing
Friday, July 4, 1997
Landing at 10:07 a.m.
Pacific Daylight Time**

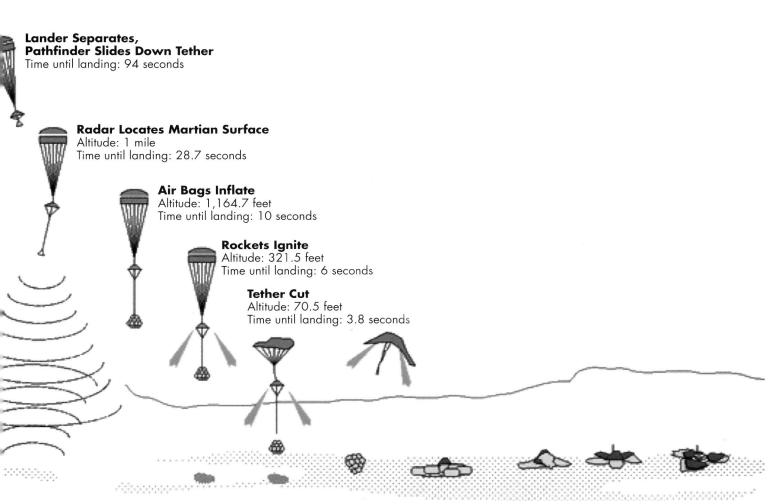

**Lander Separates,
Pathfinder Slides Down Tether**
Time until landing: 94 seconds

Radar Locates Martian Surface
Altitude: 1 mile
Time until landing: 28.7 seconds

Air Bags Inflate
Altitude: 1,164.7 feet
Time until landing: 10 seconds

Rockets Ignite
Altitude: 321.5 feet
Time until landing: 6 seconds

Tether Cut
Altitude: 70.5 feet
Time until landing: 3.8 seconds

**Air Bags Deflate,
Solar Panels Open**
Time after landing:
20 minutes

Air Bags Retract
Time after landing:
74 minutes

**Air Bags Fully
Retracted**
Time after landing:
87 minutes

The lander, bundled in its air bags, slammed into the red Martian dust at 30 miles per hour. It bounced 16 times — as high as 40 feet — like a giant beach ball. Then it rolled and rolled.

As it tumbled, a faint radio signal drifted in and out. The signal gained strength as Pathfinder came to a halt. The lander had survived its collision with Mars.

It was Sol 1. From then on, the mission would be counted in sols — Martian days — which last 24 hours and 37 minutes.

In the Mission Operations room at the Jet Propulsion Laboratory in Pasadena, hung with banners proclaiming "Mars or Bust," the Pathfinder team broke into applause. They yelled, cheered, shook hands and hugged each other. Donna Shirley, her dream realized, got tears in her eyes in the middle of an interview with a television reporter.

After their initial rush of excitement, flight controllers wondered what position Pathfinder had touched down in. If it had landed upside down, it would have to right itself by extending a petal and tipping itself over. But the lander signaled a thumbs-up to Earth. Not only had the spacecraft's unconventional belly-flop from the heavens succeeded, but Pathfinder had landed right side up.

There were no cameras on Mars to record the landing, but this painting shows the Pathfinder lander bouncing along the rocky plain of the Ares Vallis. The tether that connected the parachute to the lander was cut seconds before impact to prevent the parachute from draping, tent-like, over the top of Pathfinder.

An air bag billows over the edge of the lander's petal, temporarily blocking Sojourner's exit.

The flight team was delighted, but geologist Dr. Matthew Golombek was worried about the smooth landing. What if it was too smooth? What if Pathfinder had come down on flat ground? What if there were no rocks? The painstaking selection of a landing site, although it had involved 60 scientists from the United States and Europe back in 1994, had been based on educated guesses about the geology of Ares Vallis. There were no guarantees. Matt waited uneasily for Pathfinder to send down the first photos. It didn't take long.

Pathfinder pointed its antenna toward Earth, ready to transmit images and data. Now its camera could start taking pictures. The scientists and engineers could hardly wait. The first priority was to confirm that the air bags had retracted properly. The initial image, a black and white photo about the size of a bathroom tile, showed a section of the lander and a deflated air bag billowing like an untucked shirttail.

Members of the Pathfinder team eagerly await the first photographs of Martian terrain in more than 20 years.

As Pathfinder beamed down 2 images per minute, more of the air bags came into view. The pictures revealed that one had bunched up over the petal that Sojourner would use to exit the lander. It was blocking Sojourner's path to the Martian surface. Fortunately, Gremlin Gruel had foreseen such a problem and had a solution. Controllers told Pathfinder to lift one of its petals about 45 degrees and reel in the offending air bag.

More photos streamed down. Finally, the view of Mars beyond the lander took shape, giving the scientists what they had come for. "Rocks!" cried Matthew Golombek.

One of the first images from Mars. It's out of focus, but to the geologists it looked just fine.

After months of rehearsals in the test bed, the scientists had become accustomed to rocks hauled in from a local building supply company. Now they saw real Martian rocks. An incredible array of shapes, sizes and textures stretched out before them, dark gray in color with weathered surfaces or coatings of bright red dust.

"Looks a lot like Tucson," someone—not a geologist—commented, as he scanned the boulders and hills that jutted out in the distance.

Matt Golombek noted that the rocks leaned toward the northwest, the direction of the long ago flood. His eyes were drawn to a 10-inch tall rock. Unlike those around it, it was darker and studded with bumps that reminded Matt of barnacles. It was close to the exit ramp, within easy reach of the rover. Barnacle Bill, the science team decided, would be the first rock that Sojourner would investigate. Eventually, photos of the landing site would bloom with little yellow stickers bearing fanciful names—Yogi, Scooby Doo, Casper, Wedge, Shark, Half Dome.

It may look like a jumble of rocks, a desert without cacti or tumbleweed, but geologists can pick out clues to the Martian past by looking at shapes and textures. They can see that the small, angular rocks (blue arrow) are not as weathered as the rounded rocks (red arrows). This means that the angular rocks may have arrived at the site more recently, perhaps having been blasted from a crater formed by an asteroid collision. The origins of the white, flat rocks (white arrow) are still a mystery.

Sojourner's radio could transmit only as far as a walkie-talkie, so she could not communicate directly with Earth. Instead, Sojourner talked to the lander and the lander talked to Earth. Transmissions were made 2 or 3 times each day during the approximately 12 hours the Earth appeared above the horizon on Mars.

On Sol 2, Sojourner was ready to leave the lander. Explosive bolts blew off the cables securing her to the lander petal and 3-foot long exit ramps extended onto the surface. The rover gathered herself up to her full 12-inch height and hoisted her antenna.

Sojourner's big moment was at hand. The little rover slowly drove backwards down the rear ramp and then rolled onto Martian soil, imprints from her wheel treads trailing behind. Shouts and cheers rang out in the Mission Operations room as the lander beamed to Earth the first pictures of the rover on Mars.

On Sol 3, Sojourner got to work. Rover driver Brian Cooper adjusted his goggles and gave her the command to swing her rear end to the left and then back toward Barnacle Bill. Although Sojourner only had to travel a short distance, she had to be in just the right position. Her APXS instrument had to press its 2-inch nozzle securely against the rock in order to do a successful analysis of its chemical content. Bullseye! The APXS fitted perfectly against Barnacle Bill on the first try.

No trumpet fanfare announced Sojourner's arrival on Mars, just the whirring of her motors. Presumably, no one was there to hear it.

This is what Brian Cooper could see on his screen as he guided Sojourner toward Barnacle Bill.

The extraordinary pictures from the lander's camera helped Brian Cooper to guide Sojourner. From its perch 5 feet above the ground, the camera rotated 360 degrees, taking detailed shots of the entire landing site in black and white and color. Because Pathfinder's camera was stereoscopic, the pictures appeared 3-D. These images were used to construct a "virtual reality" map of the landing site. Looking at the map through his goggles, Brian could sense depth and pick out possible hazards. He could even "soar" like a bird above the landing site to check rock sizes, shapes and the distances between them.

Within minutes of receiving photographs, the science team posted them on the Internet via web sites that the Jet Propulsion Laboratory had launched. This was very unusual. Scientists typically wait to publish their data until after they have interpreted it themselves. In keeping with the spirit of the mission, though, everyone agreed to release the photos immediately.

On Sol 4, 46 million people logged onto the web sites, many wearing cardboard 3-D glasses to see the 3-D views. During the mission, they could browse through daily operations updates, check the latest Martian weather reports, view a photograph of sunrise on Mars or chat live with a Mars Pathfinder team member. A "web cam" located in the Mission Operations room even allowed Internet users to eavesdrop on the team at work.

On Sol 6, Sojourner left Barnacle Bill and drove to a much bigger rock about 15 feet from the lander. It would be a more eventful run-in than anyone anticipated.

Geologists had named the rock Yogi because its shape reminded them of a certain cartoon bear. At 3 feet tall, Yogi towered over the rover. Unintimidated, Sojourner, her wheels encrusted with Martian soil, approached Yogi and set about her work. She tested the soil with her APXS, then focused her camera on Yogi's pitted surface.

Sojourner's cameras capture a rover's eye-view of a looming Yogi.

The little rover's collision with Yogi was the first interplanetary fender bender.

The "virtual reality" map of the landing site was a big help when it came time for Sojourner to place her APXS against Yogi. The 3-D, bird's-eye view warned the rover team that Yogi's face, which had appeared flat, actually curved inward. Aiming for the middle would cause Sojourner to bump her head on the protruding edge of the rock, possibly damaging her solar panel or antenna. The rover would have to be aimed slightly to the left.

An alternate rover driver had the tricky job that day of judging exactly how far to tell Sojourner to turn and back up in order for the APXS to make contact. His estimate was off by a little bit and the rover ran aground, her left rear wheel riding up on Yogi's cold shoulder. Sojourner halted immediately, as she had been programmed to do, and stayed there, hanging on the rock, patiently awaiting instructions.

Guiding a rover from a computer screen 119 million miles away is a complex task. Accidents were expected. During months of rehearsals in the test bed, Brian Cooper had practiced maneuvers to free Sojourner from just such a predicament. This was the real thing.

Under Brian's guidance, Sojourner slowly backed off Yogi. Her rear wheel dropped to the ground, but there was no damage to the sturdy little rover. Now Brian had to reposition the APXS so that it made contact with Yogi. He instructed Sojourner to make one turn to the left, then another to the right. Slowly Sojourner extended her APXS and gently touched Yogi.

Success!

Until now, every move Sojourner made had been carefully plotted by its drivers. On Sol 12, Sojourner made her first try at independent navigation. Brian directed Sojourner to use her hazard avoidance systems to drive herself part way to the next target, a flat, white rock called Scooby Doo. She did it easily.

On Sol 20, Sojourner became even more independent when she made her way to a rock called Soufflé. This time, she not only had to use her hazard avoidance systems to find her way, she had to locate Souffle by herself. Using her lasers, she made it to the center of the rock, right on target.

Sojourner settled into a routine. It always began with a song that her controllers on Earth played to start their work day. This was in keeping with the tradition of sending musical wake-up calls to astronauts. There was a different song each day. "Final Frontier," the theme from the television program, "Mad About You," was the first.

Sojourner was awakened and her computer was turned on. She sent to Earth the APXS readings she had collected overnight. She updated her position. Then she was ready to rove.

Sojourner's mission was to help scientists learn about the soil as well as the rocks on Mars. By spinning her spiky wheels, the rover could expose patches of soil, and photograph the layers underneath.

By Sol 27 (July 31, 1997), the Pathfinder lander had met the goals set for its one-month mission. The mission team announced that the lander would shut down each night to save electricity. Sojourner would continue to investigate as many additional rocks as possible with her APXS.

Her designers had hoped that Sojourner would survive for at least a week in the unforgiving cold of Mars. By Sol 27, she had lasted almost 3 weeks longer than that, earning the title "the little engine that could."

Sojourner began to travel greater distances. On Sol 30, she set out for the Rock Garden. Much of her remaining time was spent navigating this imposing cluster of boulders and smaller rocks. The Rock Garden had interested scientists from the start because its smooth, angular rocks were relatively dust free and uncontaminated by the atmosphere. This would give the APXS the purest reading of their chemical content.

Sojourner makes her way through the Rock Garden.

Sojourner's camera gave scientists a series of closeup views of the Rock Garden that helped to explain the origin of many of its rocks. Rounded, pebble-like features set in a mixture of clay, silt and sand on Half Dome (top left) and Shark (top right) suggest that they were formed by deposits from the great floods. Sandblasting by strong winds on Mars is thought to have caused the pitting and fluting on Moe (bottom left).

The next target was a Rock Garden specimen called Shark. Reaching it would be one of Sojourner's toughest assignments. While a counterclockwise path around the lander would have been the most direct, it was too rocky. Brian Cooper had decided that the little rover should follow a longer and only slightly less perilous clockwise trail. Threading through the maze of rocks that formed the gateway to Rock Garden proved so tricky that Brian and his team dubbed it the Bermuda Triangle. Several times the rover's hazard avoidance systems turned her off automatically to prevent spills. To make matters worse, her gyroscope had not been working properly. It now caused her to drift so badly that Brian finally turned it off.

Sojourner reached Shark on Sol 52. Six days later, the rover's backup batteries ran down. The rover team had to adjust Sojourner's schedule so that she did not start her work until the sun was up high enough to provide sufficient power to run her computer. From then on, the rover could only take APXS readings during the day. Still, she managed to complete readings of Half Dome and Chimp.

The wind has played a leading role in sculpting the Martian surface. Wind also formed these dunes that Sojourner discovered tucked behind the Rock Garden.

After finishing her exploration of the Rock Garden, Sojourner was to head back to Pathfinder. The team would never know whether she made it.

The lander sent its last complete transmission to Earth on Sol 83 (September 27, 1997). When the operations team tried to set up their usual communications link on the following day, there was no reply from Pathfinder. The team was able to lock onto a signal from Pathfinder's auxiliary transmitter on October 1, which indicated that the spacecraft was still functioning. There was one final blip on October 7, and then nothing more.

The team thought that, as expected, the lander's battery had finally run down. This knocked out the clock that told Pathfinder when to point its antenna toward Earth to receive the command to turn on its transmitter. With the batteries dead, the lander could no longer run heaters at night. The lander's electronics eventually froze.

After 5 days without word from Pathfinder, Sojourner's computer program would have told her to return to the lander. Still waking each morning, she would have circled Pathfinder at a distance of about 10 feet. She would have repeatedly asked the lander for commands from Earth, but there would have been no reply.

On November 4, 1997, the Mars Pathfinder team finally and reluctantly announced the end of the mission. They would still try once a month to pick up a signal from the lander, but they were not successful.

The team's sadness was tempered with joy at its achievements. It was less a "so long" than a "see you later," because, thanks to Pathfinder, more missions to Mars lay ahead.

Pathfinder had demonstrated that a lander could be parachuted safely onto Mars, that a microrover is an effective vehicle to explore another planet, and that space exploration can be done quickly and relatively cheaply. The mission had cost $266 million, exactly what had been budgeted.

Sojourner had only driven around an area the size of a big back yard, but she had gathered a wealth of valuable scientific data. The mission returned 2.6 billion bits of information, including 16,000 images from the lander and 550 from Sojourner, 15 thorough chemical analyses of rocks and soil from Sojourner's APXS, and millions of readings of Martian temperature, pressure and wind.

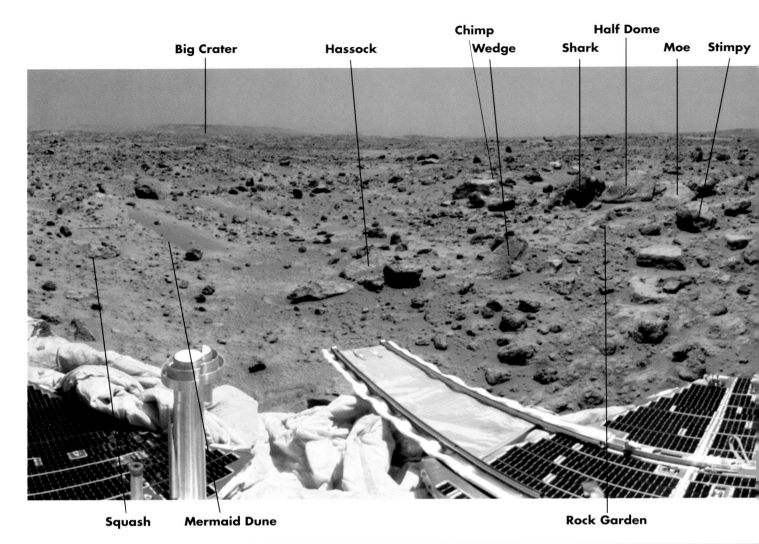

Big Crater **Hassock** **Chimp** **Wedge** **Shark** **Half Dome** **Moe** **Stimpy**

Squash **Mermaid Dune** **Rock Garden**

Yogi **Lamb** **Cabbage Patch** **Casper** **Scooby Doo**

Twin Peaks   **Photometry Flats**   **Cradle**

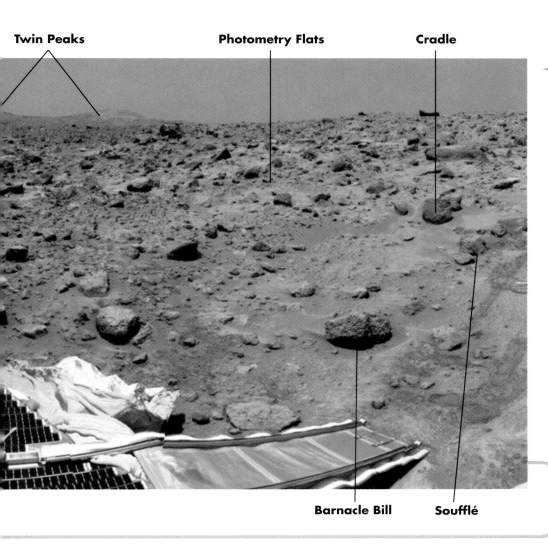

This panorama was made up of thousands of images taken by the lander camera over the course of the mission. The camera photographed a different section of the landing site at the same time each day to ensure that the lighting would be consistent throughout.

Barnacle Bill   **Soufflé**

Roadrunner Flats

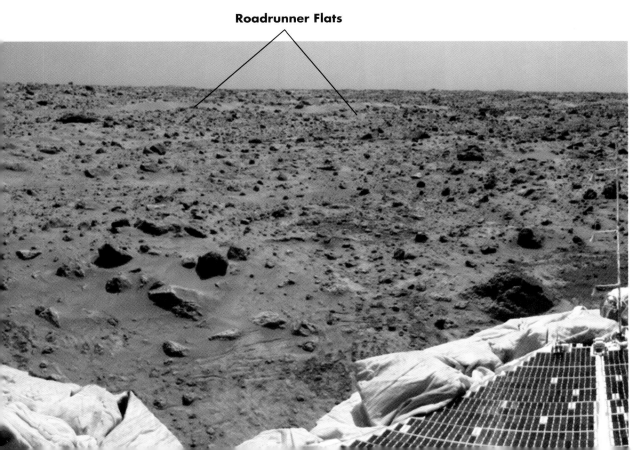

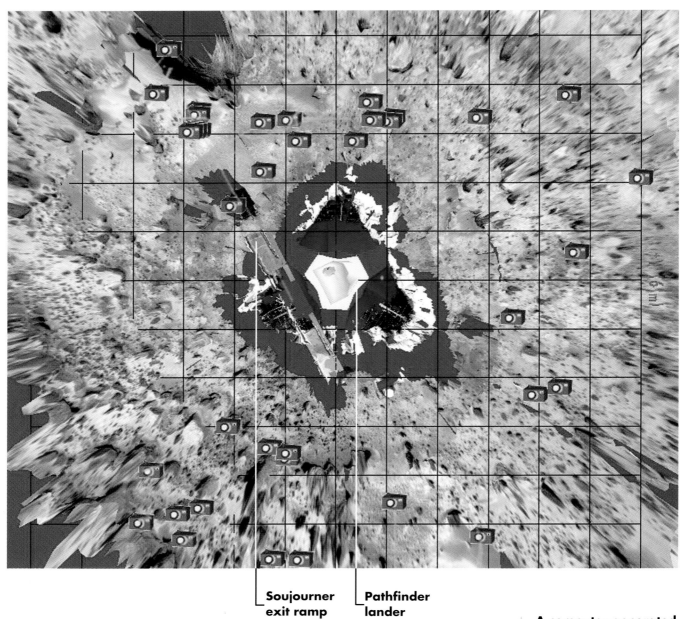

Soujourner
exit ramp

Pathfinder
lander

A computer-generated overhead map shows Sojourner's path during the mission. Each camera icon indicates where she stopped to take photographs.

The mission confirmed that liquid water had once flowed through the Ares Vallis and that the planet had been shaped by a warmer atmosphere that produced clouds, winds and seasonal changes. Mars had once had conditions to support life.

Sojourner's APXS measurements of the Martian rocks revealed surprises. The geologists expected the mineral content to resemble that of meteorites from Mars. But some of Sojourner's targets, including Barnacle Bill and Shark, had more in common with Earth rocks. The scientists do not know why.

Only additional exploration will reveal the answer to this and other questions about Mars — including whether life had existed there in the past or even exists there now.

Pathfinder lander photographed by Soujourner.

Beginning in 2001,
Mars Surveyor
orbiters will provide
communications links
between Mars and
Earth for future rovers
and landers. This will
enable rovers to cover
more Martian
territory in pursuit of
rock and soil samples.

A multi-year program of robotic exploration called Mars Surveyor has already begun. It will launch 2 spacecraft to the Red Planet every 26 months.

The next generation of rovers will be even smarter and more durable. Three to 5 times the size of Sojourner, they will generate more solar power to travel longer distances at greater speeds.

These rovers will have more ambitious scientific goals, too. They will carry robotic drills for boring into rocks and collecting small core samples. These will be stored by the rovers in sealed containers to be picked up by later missions. Before being returned to Earth, the samples will be analyzed by tiny, magnifying cameras and sophisticated spectrometers.

This model called Rocky 7 shows some of the features that might be found on future rovers. A robotic arm would carry instruments for gathering samples of Martian rocks and soil, storing them for later pickup.

Scientists hope the rovers' instruments will allow them to select rock samples that might contain evidence that life once existed on Mars. The earlier missions in search of life on Mars analyzed only the top layer of soil, which has been exposed for eons to killing cold, ultraviolet radiation and cosmic rays. Some scientists believe that life, perhaps in a form unknown on Earth, may yet be found inside the rocks or at greater depths below the surface.

The Mars Surveyor missions will set the stage for human exploration. Experts believe that the first human missions to Mars could depart as early as the first decades of the 21st century.

In the distant future, people from Earth might live in colonies on Mars in special shelters protected from cold and radiation. They might extract fuel from the carbon dioxide in the Martian atmosphere, draw water from geothermal pools beneath the surface, manufacture bricks and glass from the soil, and grow food in inflatable greenhouses. The descendants of Sojourner, rovers the size of small campers, could be the vehicles of choice for human travel on Mars.

The Pathfinder lander was renamed the Carl Sagan Memorial Station for the much honored astronomer who had worked on several Mars missions and who encouraged public interest in space exploration. The Sagan Memorial Station and the Sojourner rover remain on Mars.

Perhaps, one day in the future, a colonist driving a fancy new rover might happen upon Sojourner, half buried in a drift by the Martian winds, her small antenna still upright. Maybe the colonist will linger on the lonely plain, tracing her name in the red dust blanketing Sojourner's long dead solar panel, and remember the adventure that started it all.

The mission to retrieve samples from earlier flights is scheduled for launch in 2005. Scientists will finally be able to examine Martian soil and rocks first-hand for signs of past or present life.

THE ROVER TEAM AND FRIEND

1960

Koralb 4 [USSR] Failed to leave Earth
Koralb 5 [USSR] Failed to leave Earth

1962

Koralb 11 [USSR] Blew up
Mars 1 [USSR] First probe to pass Mars [120,000 miles] • Contact lost

1964

Mariner 3 [USA] Went into wrong orbit and missed Mars
Mariner 4 [USA] First successful mission to pass Mars • 22 photos of Mars surface
Zond 2 [USSR] Passed Mars by less than 1,000 miles • Communication failed

1965

Zond 3 [USSR] Mars orbit • 25 pictures of the far side of Martian moon

1969

Mariner 6 [USA] Passed Mars at 2,100 miles • 75 photos and data
Mariner 7 [USA] Passed Mars at 2,200 miles • 126 photos and data
Unnamed 1 [USSR] Failed to leave Earth
Unnamed 2 [USSR] Failed to leave Earth

1971

Mariner 8 [USA] Fell into Atlantic
Cosmos 419 (USSR) Orbiter/Lander • Failed to leave Earth
Mars 2 [USSR] Orbiter/Lander • Successful orbit, but few clear pictures because of Martian dust storm • Lander crashed
Mars 3 [USSR] Orbiter/Lander • First soft landing, but failed after 2 minutes • Small portion of one photograph
Mariner 9 [USA] Successful orbit • Over 7,000 photographs of entire globe • Provided first atlas of Mars

1973

Mars 4 [USSR] Orbiter • Failed to orbit • Passed planet
Mars 5 [USSR] Orbiter • Orbited for 10 days before failing • Returned some photos
Mars 6 [USSR] Orbiter/Lander • Lander crashed on Mars
Mars 7 [USSR] Orbiter/Lander • Missed Mars

1975

Viking 1 [USA] Orbiter/Lander • Successful orbit and soft landing
Viking 2 [USA] Orbiter/Lander • Successful orbit and soft landing
Combined total of 52,000 images from orbiters and 4,500 from landers

1988

Phobos 1 [USSR] Orbiter/Lander • Investigates Martian moon (Phobos) • Lost
Phobos 2 [USSR] Orbiter/Lander • Investigates Martian moon • Reached Martian orbit • Photos of Mars and Phobos

1992

Mars Observer [USA] Orbiter • Returned some images as it approached Mars • Lost

1996

**Mars Global
Surveyor** [USA] Orbiter • Successfully entered Mars orbit September, 1997
Mars 96 [RUSSIA] Orbiter and 4 landers • Failed during launch

1998

**Mars Climate
Orbiter** [USA] Will study Martian climate

1999

**Mars Polar
Lander** [USA] Will study soil and weather near Martian South Pole

2001

**Mars Surveyor
2001 Orbiter** [USA] Will serve as communications link for following lander

**Mars Surveyor
2001 Lander** [USA] Will carry a rover that will spend a year drilling for core samples of Mars rocks

2003

**Mars Surveyor
2003 Orbiter** [USA] Will serve as communications link for following lander

**Mars Surveyor
2003 Lander** [USA] Will carry an advanced rover and perform further studies of Martian rocks and soil

2005

**Mars Surveyor
2005** [USA] Will return to Earth with soil and core samples from previous landers

2018

Human Mission

	Mars	**Earth**
Diameter	4,222 miles	7,926 miles
Average distance from sun	142 million miles	93 million miles
Single revolution around sun	687 Earth days	365 Earth days
1 day	24 hours, 37 minutes	24 hours
Moons	2	1
Atmosphere	Carbon dioxide, nitrogen, trace oxygen, argon	Nitrogen, oxygen
Polar caps	2	2
Magnetic field	None detected	Yes
Weight of 100 pound barbell	38 pounds	100 pounds
Seasons	Yes	Yes
Highest temperature	63° F	136° F
Lowest temperature	-199° F	-128° F
Life	Maybe	Yes

www.mars

There are hundreds of Mars web sites. These 3 sites will provide links to the best of them.

The Jet Propulsion Laboratory's past, present and future missions to Mars and other planets.

http://www.jpl.nasa.gov/

Images of Mars and other planets.

http://photojournal.jpl.nasa.gov/

NASA

http://www.nasa.gov/

Photo by Eric Wunsch

Susi Trautmann Wunsch is a freelance journalist. She lives in New York with her husband and their two sons. This is her first childrens book.